Socks Are Like Pants, Cats Are Like Dogs

Games, Puzzles, & Activities for choosing, identifying, & sorting math

Creative Commons Attribution-NonCommercial-ShareAlike license
by Natural Math®, Malke Rosenfeld, and Gordon Hamilton

ISBN: 978-0-9776939-0-0 (print)
ISBN: 978-0-9776939-4-8 (ebook)
Library of Congress Control Number: 2015958237

Socks Are Like Pants, Cats Are Like Dogs: Games, Puzzles, and Activities for Choosing, Identifying, and Sorting Math

Illustrations by Gordon Hamilton and Ever Salazar
Beetle photographs by Udo Schmidt, used with permission
Cover by Mark Gonya
Edited by Maria Droujkova and Carol Cross
Layout by Jana Rade

Published by Delta Stream Media, an imprint of Natural Math®
309 Silvercliff Trail, Cary, NC 27513
January 2016

This book came about thanks to parents, teachers, and math circle leaders who tested the activities, gave feedback, and crowdfunded the production of the book. Thank you for your support!

Anonymous
Alana Schoepp
Alexander Shpilman
Alexandr Rozenfeld
Alexandra Ovetsky Fradkin
Algot Runeman
Ali Ghasempouri
Benji Jasik
Brandy Wiegers
Brian Stockus
Cameron Goble
Carolyn Galbraith
Catriona Kaminski
Charles Settles
Charles Tysoe
Christopher Danielson
Daniel Finkel
Daniel J. Quinto
David Wees
Deborah Myerson
Dee Mosley
Denise Gaskins
Derek Fok
Donna Rund
Elena Koldertsova
Elena Malkina
Elizaveta Shevyakhova

Emily Kolatch
Fee Kapadia
Gerald P. Ardito
Heather Haines
Javier Garcia
Jenny Allen
Jeremy
Joel Tornatore
Jonathan Edmonds
Joshua Zucker
Judy Keeney
Julia Brodsky
Julie Secrest
Karla Momberger
Lant
Katherine Kearns
Kenny Felder
Lhianna Bodiford
Loren Freed
Luke A. Sinden
Mackedie Spiker
Maria
Marina Pratchett
Marisa Harder-Chapman
Mark Trushkowsky
Martha Steel
Marmouze
Matthew J. Olwell
Megan Gibbs

Megan Schmidt
Michael Jacobs
Michelle Garrigan
Miranda Jubb
Mona Hennigar
Nancy McCullen
Natalia Shif
Natalya Averina
Oliver John Golden
Paula Krieg
Richard Kingston
Richard Rosenfeld
Roberta Kerler
Sandra
Sandra Kozintseva
Scot B. Bailey
Sian Zelbo
Simon Gregg
Simon Terrell
Sophie Haroutunian-Gordon
Stephanie Johnston
Sue VanHattum
Susan
Susan Paulus
Tatiana Marquardt
Vanita
Yelena McManaman
Zinovy Shekhtman

CONTENTS

INTRO

As parents and teachers of young children, we know that kids are experts at noticing. We also know that kids *love* to talk about what they see. These kinds of conversations can happen anywhere—at the grocery store, at the park, on a walk, looking at a book, or on car rides. And, many times, these conversations are ripe with potential for finding and talking about math.

A light-hearted conversation about what children (or parents!) notice is a perfect opportunity to help kids use and develop core thinking skills needed in making math. For example, two socks are the same because they're both socks. The two socks are *not* the same because one is solid red and the other has polka dots. Socks and pants are the same because they're both items of clothing and they come in pairs! But they're *not* the same because socks are (usually) for feet and pants are (usually) for legs. These kind of conversations help build a mathematical understanding of **equivalence,** meaning that one thing can be like another if you focus on an aspect with similarity or sameness.

You can help your children deepen their abilities to notice similarities, sameness, and differences with questions like these...

How is this dog different from that one? How are they the same?

You have two socks on, but there's something that's different about them—can you figure out what it is?

...and listen closely to the answer. You may hear something that sounds silly or profound or completely out of the box. The best part is that you are free to find as many answers as you like as long as they make sense to you. And, with each answer, you get a precious glimpse into the inner workings of your children's minds and how *they* see the world.

Conversations like these help children learn to describe the properties, or attributes, of what makes up the object or person in question. The beauty of identifying attributes is that there are endless combinations of similarities, sameness, and differences to be found in the world. And, there is power and joy to be found in making your own rules for sorting that follow your interests.

For example, your daughter has all her stuffed kitties around her in a pile. How many white kitties are there? What if a kitty has some orange in her fur along with white? Is white/orange a new category? This kind of activity and ensuing conversations is centered on the math idea of *categorical variables* and the process of sorting things by category. Categories are groupings of items or ideas with similar attributes; the categories can change depending on what you are trying to discover. Your daughter might not be interested in sorting the kitties by color, and instead sort them into piles of sick and well. The kitties don't change, but the categories do.

Or say you and your son are baking a cake. Your recipe calls for one cup of sugar. You might choose to halve that amount, but your child wants to double it! That's a variable. Which will you choose? Which will your son choose? Variables are like that. They demand that we make a choice, and each choice has an impact on the final product.

Identifying **attributes** (particular qualities with which you can describe an object) and *choosing* **variables** are mathematical processes that open up the world of algebraic reasoning to all who participate. As children choose from an inventory of variables to create their artwork, they experience the process of how algebraic ideas come to be. And, as children find, identify, sort, and discover attributes in games and puzzles, they grow in their abilities to notice structure, order, and pattern, all necessary skills for mathematical activity at any level. Helping kids make explicit their observations about attributes and their choices related to variables is what this book is all about.

A note about the structure of this book:
It can be challenging to recognize math ideas in natural or artistic contexts rather than simplified worksheets and textbooks. How do we help our readers find the math in the playing, making, and solving? In addition to the directions, each activity includes a **Note to the curious** and **Keywords.** "Note to the curious" are written in adult language that help you understand the mathematical thinking inherent in the design of each game, puzzle, or art project; "Keywords" can help you google additional information.

We hope you and your children will thoroughly enjoy solving, sorting, making, and identifying your way through the activities, games, and projects in this book!

~Malke & Gord!

Note to the curious:
Things can be connected in multiple ways, but it can be challenging to find good connections. These puzzles are only as strong as their weakest link.

Keywords: Sorting, categorizing, graph theory, attributes

THIS IS LIKE THAT

This game creates a chain of association between seemingly unrelated objects. Look at each object in the puzzle and place them in the circles so that objects in connected circles share a common trait.

Once everyone has figured out in their own heads what the connections are, share yours with the group. You might be surprised how different they are! Our own solution might be one of many possible solutions.

The whole family can be engaged in solving these puzzles. You can also make your own game variations without the actual puzzles. Suggestions for variations follow the puzzles. Have fun!

egg - piggy bank - fried egg - pig - chicken - euro coin

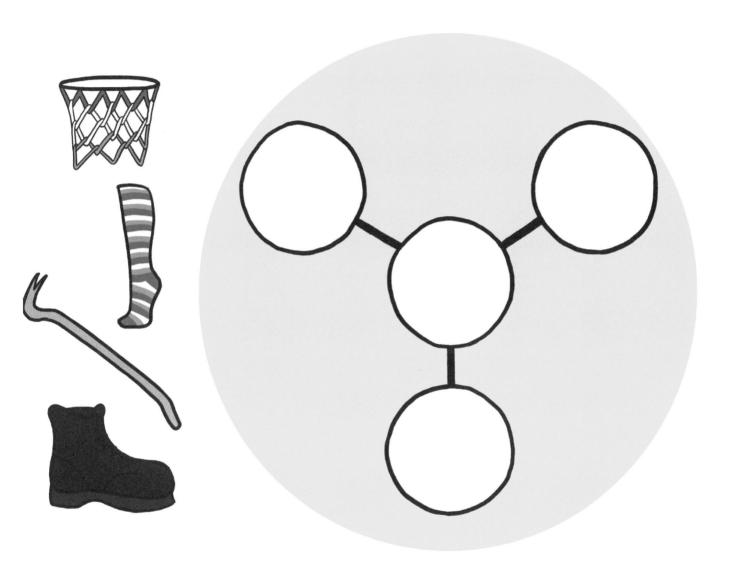

basketball net - sock - crowbar - boot

11

cow - T-shirt - glass of milk - glass of juice - leather jacket

bananas - baseball glove - mortar & pestle - curling stone

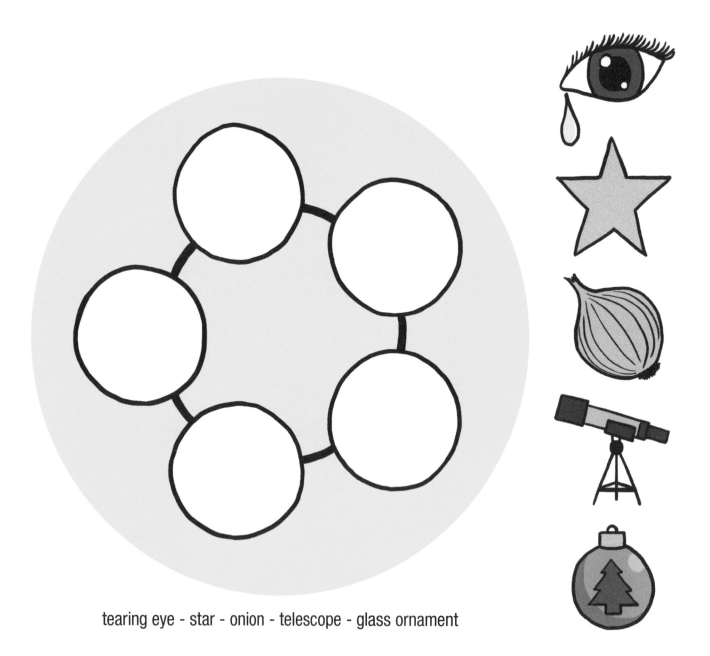

tearing eye - star - onion - telescope - glass ornament

14

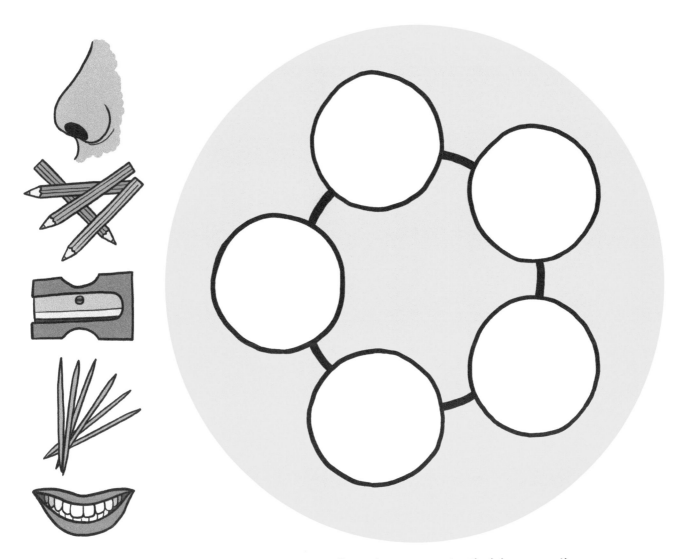

nose - pencils - sharpener - toothpicks - mouth

Game Variations

Chain of Attributes (Ages 4+)
Start with an object. Take turns looking through the house for a new object that can be added to the last object at the end of the sequence. The added object must have something in common with the previous object and must share an attribute not previously used. Repeat and grow the chain as long as you like. For example:

Jane starts with the family cat.
Paul adds a green plant (both are alive).
Dad adds a green truck (both are green).
Mom adds a red truck (both are trucks).
Jane adds a red ball (both are red).
Paul adds a glass marble (both are round).
Dad adds a glass mug (both are glass).
Mom adds the cat's metal drinking cup (both are for drinking).

After some moves it's always fun to try to return to the beginning as was done here, but this requires finding something that connects both to the start and the end of the list—not easy!

Make it Fit! (Ages 6+)
• Gather up some interesting items from around the house.
• Draw a web of circles linked by some lines, like those above.
• Choose some objects and place them in the circles.

Objects that are linked in the web must share an obvious attribute. Objects that are not linked in the web must not share an obvious attribute. Different players will see different attributes, so don't expect everybody's answer to be the same. Choose your objects carefully—otherwise they'll be impossible to place in your web.

The Missing Link (Ages 6+)
In this puzzle a player skips a connection which another player must find before contributing their own. For example:

Jane starts with the family cat and the three-legged table. What is the missing link between the cat and the three-legged table?

Dad guesses that the missing link was the four-legged table (which is connected to the cat because of the number of legs, and the other table because they were both tables). Dad must now extend the list from the three-legged table. He thinks of a new object: the family sled. The family must find the missing link between the three-legged table and the family sled.

Mom thinks for a moment and says "tricycle," which connects the three-legged table to the family sled.

Record your results! All these activities produce puzzles that can be shuffled up and given to other people. You can also play this on a car trip.

Around a Star (Ages 8+)
Solve "Around a Star" puzzle by placing a word or picture at the end of each spike so that neighboring spikes share an attribute. As you create your own "Around a Star" puzzle, your stars may have few or many spikes. When they have a lot, you'll need to fill some of the vertices to keep the difficulty to a manageable level. You can find answers on page 58.

Around a Star 1

selfish - fish - sing - caring - bass

Around a Star 2

octopus - liquid - october - squid - fall - spill

Around a Star 3

tooth - teapot - leaves - dental floss - reptile skin - snake - fern - violin - snowflake - bell

BEADING PATTERNS

We often ask children to find and sort everyday objects according to their properties (for example, into piles of white socks and socks that are not white). In early algebra terms, this means sorting items into categories. We can also use this kind of mathematical thinking about properties to create brand new objects.

Beads are full of similarities and differences that can help you create beautiful patterns as simple or complex as you want. All you need to get started are string, pipe cleaners, and beads in multiple colors and sizes. You can even make your own beads by cutting up drinking straws! As you create your patterns, you get to ask lots of interesting questions:

"What beads will help me make a pattern that is interesting to me?"

"Do I want to use all the small beads, or a combination of sizes?"

"Do I want all my beads to be smooth, or should I add a rough textured bead into the mix?"

Whatever you choose to do, it will be yours and it will be beautiful. You will know exactly how to talk about what and how you made this beautiful thing because you're the one who created it!

Note to the curious:
An object can be made of smaller, simpler units, such as a necklace out of beads. Being able to see a composite unit or a pattern as a whole can be challenging for young learners. Making beaded patterns and describing the individual beads using categories such as color, size, and shape can help children begin to see and repeat the salient details in their pattern.

Keywords:
categorical variable, numerical variable, units, grouping, multiples, composition of units, properties, functionality, functional relationships, observing/ discovering categories

MATERIALS

- *Pre-cut lengths of cotton string and/or pipe cleaners*
- *Plastic pony beads, wooden beads, and/or beads made from plastic straws*
- *Bowls or other containers to hold beads*

ACTIVITY DESCRIPTION

Look at all the beads. Find different ways to describe them (color, shape, size, texture, etc.). Notice the similarities and differences between the beads (for example, same color, different shapes). Create a four-bead pattern unit and repeat that unit until satisfied.

Adaptations by Ages

Babies
Use very large beads or other objects like balls or blocks. Have baby handle and play with objects. Comment on their texture, shape, and color while baby is playing. Line up objects on the floor in front of baby to create a short pattern; repeat pattern one or two more times. Point to each bead and name one attribute category at a time (for example, *"smooth, rough, smooth, rough..."* then *"red, blue, red, blue..."*).

Toddlers
Provide pipe cleaners and a selection of large wooden beads. Let your child experience the beads by touching and stringing them, but don't worry about patterns for now. Talk through your own making process while your child makes hers alongside you. Talk about why the bead you are using is different from (or the same as) the one your child is using or about what comes next in your pattern.

Older Kids
Use an interesting assortment of beads and pipe cleaners (or string), three or more attributes (such as bead shape, color, and texture), and three or four beads to make pattern unit. Make your own alongside your child. Take turns investigating each others' work—how is your child's pattern similar to yours?

BEETLE SORT

Bugs! Kids either love them or are repelled by them. Because of their size and the fact they can scurry away quite quickly, it's not always easy to look closely at the beauty and structure of such creatures. Here's a chance to sort through some families of beetles using the beautiful photographs of beetle collector Dr. Udo Schmidt.

In this series of puzzles, you can sort beetles based on how they look. Before the advent of genetics, this is how we categorized all animals. Sorting requires finding attributes. Some beetles have thicker legs, kinked antennae, or powerful mandibles.

Each puzzle consists of representative beetles from two families with two beetles on the left page and a whole bunch of beetles on the right page. Follow the directions on the left side and collect only the beetles that are indicated.

Encourage children to discuss why they think a beetle should be collected. Ask children to explain their reasoning. Accept all answers with explanations as possibilities. Mistakes should be expected. When working on the book, one of the authors (Dr. Gordon Hamilton) solved two of the puzzles wrong, at least according to the current scientific classification of beetles in the answer keys. Free play on their own terms helps children feel good about math. Toward that goal, children can arrange beautiful beetles in their own ways. On the other hand, tenacity in the face of failure also protects against math anxiety. To build up tenacity, help kids to figure out how the scientific classification works. If your child is getting frustrated, blame the beetles! It's their fault the puzzle is so difficult!

You'll find the beetle puzzles on the next page, and plenty more at the back of the book. Ready... Set... GO!

Beauty and mathematics are inseparable.

In this puzzle we use the beauty of beetles to inspire K-2 students to find and discuss patterns.

Note to the curious: Too often the sorting jobs we give our children are not very challenging. Their young brains are capable of differentiating complex patterns like those of identifying beetle families. Let them flex these sorting muscles!

Keywords: Sorting, categorizing, traits, attributes

Collect 3 beetles from the
Tenebrionidae family.

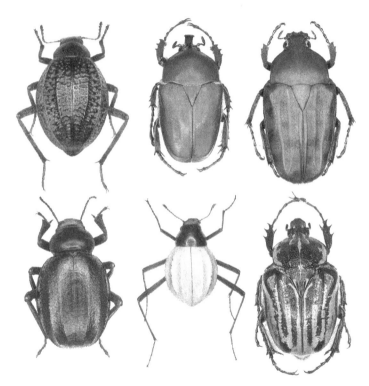

Collect beetles from the
Chrysomelidae family.

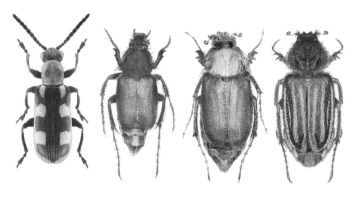

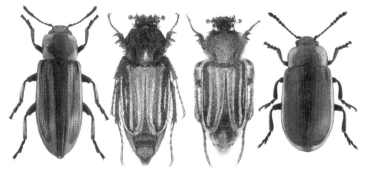

Avoid those from the
Glaphyridae family.

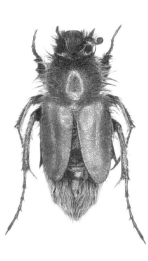

28

Collect beetles from the
Staphylinidae family.

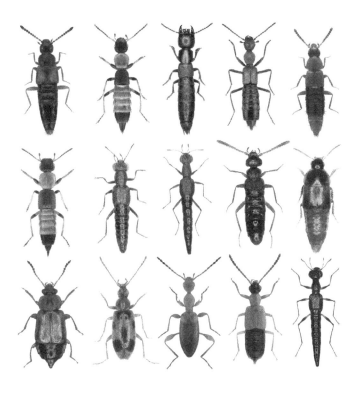

There are 2 impostors from
the Anthicidae family.

Collect 3 beetles from the
Colydiidae family.

Avoid those from the
Geotrupidae family.

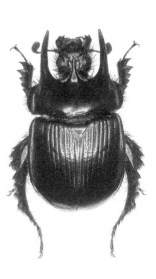

30

Collect beetles from the
Cerambycidae family.

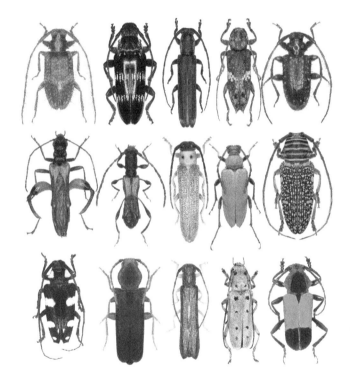

Avoid one impostor from
the Oedemeridae family.

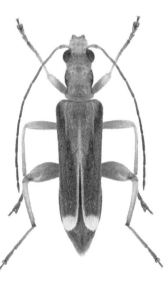

31

Collect 4 beetles from the
Cholevidae family.

Avoid those from the
Scarabaeidae family.

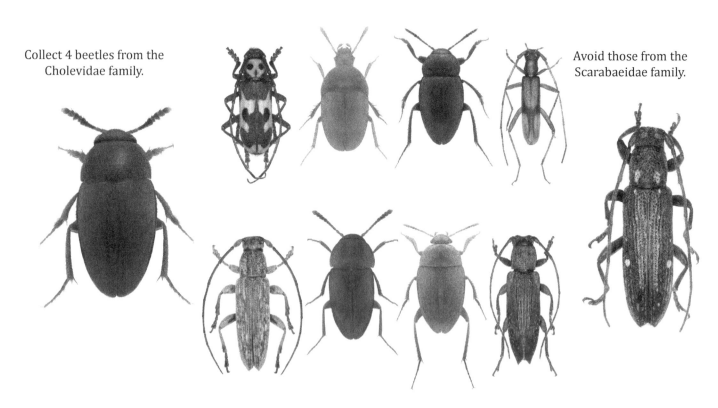

All the Same!

Making Paper Pizzas

Mmmmm... pizza! Let's make our own! It's easy—just design one piece of your perfect pizza and then make three exact duplicates, paying attention to attributes such as color, shape, and position of the paper "ingredients." The challenge for young makers is to focus on multiple attributes including number, shape, and position, all at the same time. Don't hesitate—make some pizza today!

MATERIALS

- Colorful paper circles (trace around a large bowl, cut out circle)
- Geometric shapes cut or punched from colored paper (you can use Word 'insert--shapes' to draw and duplicate shapes to cut out)
- A plain piece of paper on which to glue the pieces after you've designed them
- Glue sticks
- Scissors

ACTIVITY DESCRIPTION

Pre-cut the "pizza crust" circles and shapes before starting. Make sure that all the items in each category of shape, such as square, are the same size.

Fold and then cut the circle into four equal pieces.

Place all four pieces in a line, point down.

Choose three or four shapes and glue them down in a pleasing design on the first piece.

Repeat the same design on each piece, paying special attention to the positions of the shapes in relation to each other and the edges of the piece.

While child(ren) and adult(s) are making their together, ask clarifying questions such as: *"Does that look the same to you?"* or *"Does each piece of your pizza have the same number of shapes?"*

When satisfied with your work on each of the four pieces, glue them back into the original circle on a piece of paper, and admire the amazing results. Each slice is the same and yet the effect of putting them back together reveals an incredible new design. What do you notice?

Adaptations by Ages

Babies
Use colorful blocks to make a three- or four-block design in front of baby, talking all the while about the colors and shapes. Use spatial and positional words: *"The red block is on top of the green block."* Repeat the design and your words. And again! And again!

Toddlers
Cut the circle into fourths while your toddler watches. Work together with your child to choose shapes for the first design. This choosing process is a great time to have conversations about similarities and differences between the paper shapes. You may need to apply the glue, but allow your child to pick the position and place the shape on the piece. That is a lot of fine motor coordination for a little one. When it is time to replicate the design on the remaining three pieces, you may want to finish up or you can make it a shared process. While you work, talk out loud about what you are doing. Ask questions like *"Where does the red circle go? In the middle? Near the point?"* and encourage your child to give you directions.

Older Kids

Enjoy a shared making experience, each of you with your own paper pizza to design and make from start to finish. The four pieces should be laid out in a row point down and assembled into a circle at the end. This transformation takes a pattern with translational symmetry and creates a pattern with rotational symmetry. Throughout the activity, adults can model observations:

"I just noticed that you are using three shapes and I am using four."

"I wonder how it will look when we put all the pieces back together into a whole pizza?"

"I noticed that your green circle on this first piece is close to the edge, but on this other piece it's closer to the center. What can you do to make sure they're in the same position?"

Note to the curious:

One object can be made of many smaller, simpler units. For example, pizzas can be baked with a whole bunch of different toppings. Simple rules such as deciding on the number of "toppings" and their positions on a slice of paper pizza can lead to surprisingly complex designs.

Keywords:
sameness, position, units (pizza slice), composition of units, categorical variable, transformation, symmetry, translational symmetry, rotational symmetry

DANCING VARIABLES

Did you know that all young mammals share the same trait in their play? It's called **galumphing!** Galumphing is the exuberant use of one's body and energy at every opportunity, even in simple activities like moving down the street. Why walk when you can hop? Or skip? Or...? This dance and song is perfect for combining your little galumphers' energy into some math-y, dance-y fun!

MATERIALS

First verse of the song Rig a Jig Jig
www.youtube.com/watch?v=l7wpxjNAITg
[Words adapted here to focus on the movement
variables]

*As I was **walking** down the street, down the street,*
down the street
*As I was **walking** down the street, hi ho, hi ho, hi ho*
Rig a jig jig and away we go, away we go, away we go,
Riga a jig jig and away we go, hi ho, hi ho, hi ho

ACTIVITY DESCRIPTION

This song has a simple melody and structure
that invites movement and play. Sing (or simply
chant) this song when you are out and about
and find ways to change the word "walking" to
another verb. For example, walking can become
jumping, running, or skipping. Or, you can meow,
roar, bark, sparkle, blast off, etc.

Notice the pulse of the song (child in video
shows it naturally with her head as she sings).
This is the down beat on which you walk/jump/
leap/march/etc. You can also play around with
other variables that affect this dance activity. Try
variations in tempo (slow, medium, fast), volume
(whisper or shout!), or size of movements
(small, medium, giant).

Locomotor Movement Suggestions:

walking, jumping, leaping, marching, skipping, sliding, galloping

Note to the curious:
Identifying patterns of change is a key algebraic reasoning skill. Changing the attributes of a dance such as the rhythm, tempo, or the body part in motion changes the experience of the dancer and the look of the dance, creating different outcomes.

Keywords: categorical variable, rhythm, patterns, locomotor movement, dance, physical, moving-scale math

VENN DIAGRAMS

Venn diagrams are one of the most stunningly beautiful mathematical constructs that children should experience. Most only get up to the three-circle Venn diagram, but that should really be the starting point.

To start, populate the following Venn diagram:

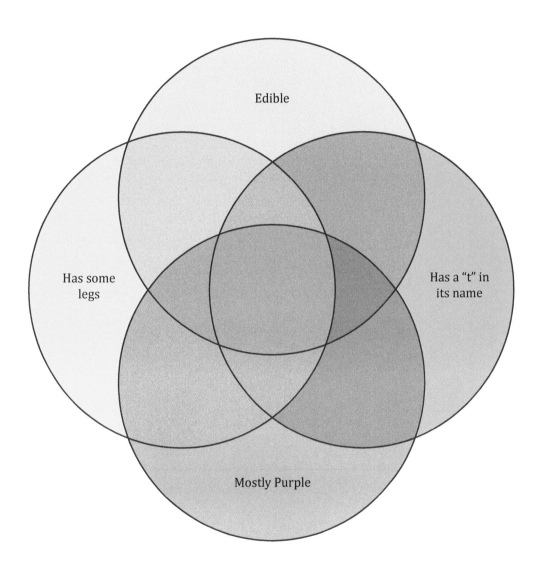

You might find that things are falling into place well... until a tiger knocks at your door! Of course, you must ask him to leave because you just have no room for him. He needs to be in BOTH the yellow and pink circles but not the others. Alas, he takes offense, and you find yourself categorized in the green and pink circles.

The diagram on page 41 is actually not a Venn diagram because a tiger and red pepper defy being categorized. Real Venn diagrams have exactly one region for every categorizable object, like the one on page 43. Fill it with people you know. It may take a year! For example, your left-handed grandmother is an adult so she needs to go in the purple ellipse. She never eats cereal so she must go outside the green ellipse.

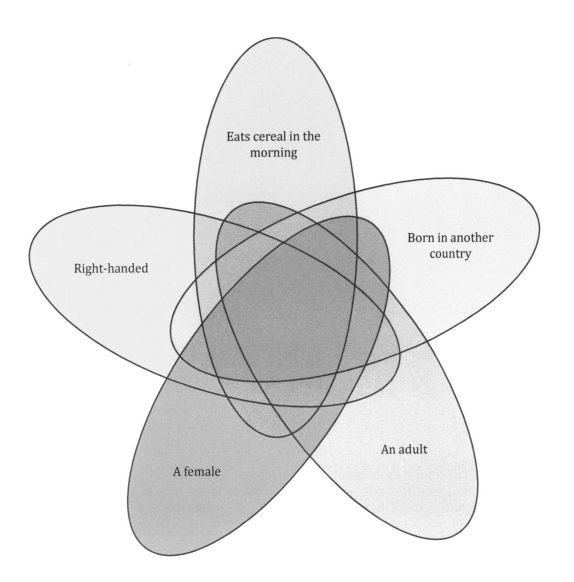

Eats cereal in the
morning

Born in another
country

Right-handed

An adult

A female

Select your own criteria in the following Venn diagrams with five and seven questions. Can you populate each region or is it impossible?

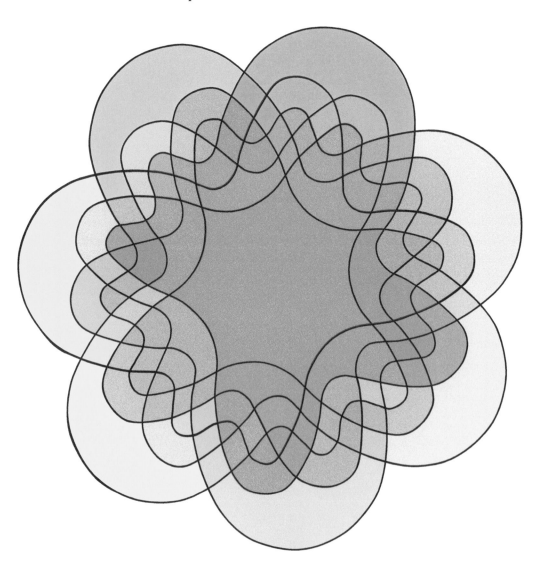

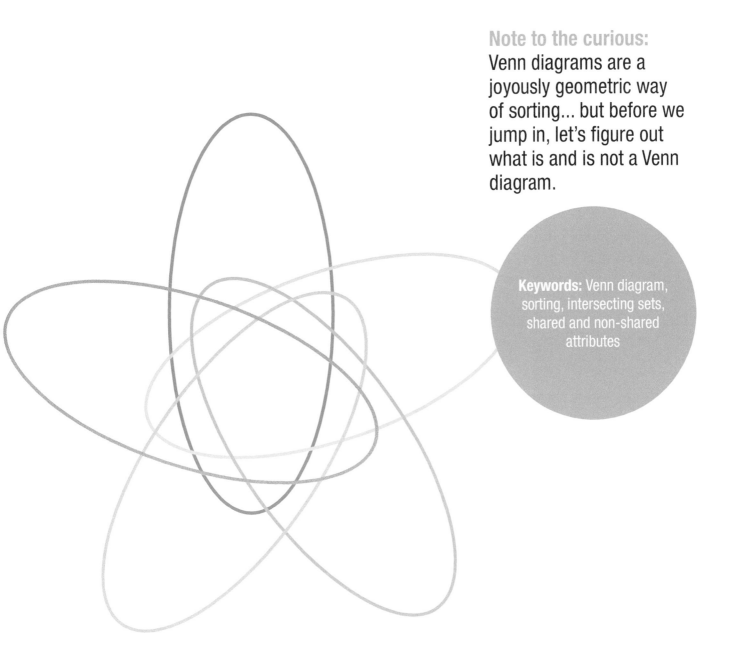

Note to the curious:
Venn diagrams are a joyously geometric way of sorting... but before we jump in, let's figure out what is and is not a Venn diagram.

Keywords: Venn diagram, sorting, intersecting sets, shared and non-shared attributes

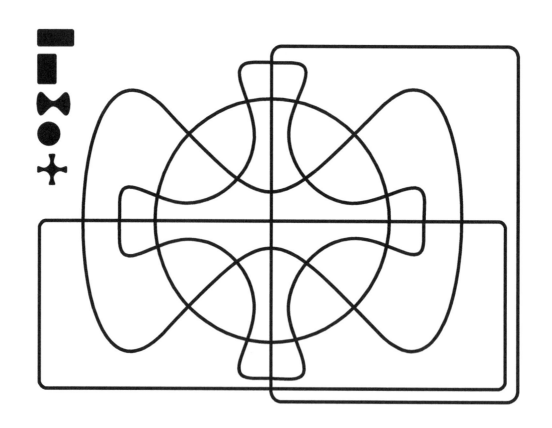

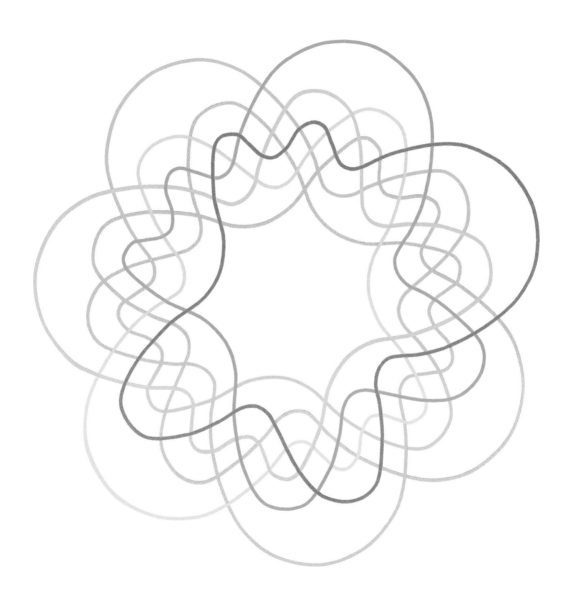

Geometric Grid Designs

Sometimes the coolest things can be made from an inventory of limited choices. For example, what if you only had right triangles and squares, two colors, and a four-cell space in which to make your design? As you experiment with different design ideas within the limits of a four-square grid, you are making active, mathematical choices about the categorical variables of position, shape, and color.

MATERIALS

- *Grids (templates can be downloaded at http://www. mathinyourfeet.com/6/ post/2013/04/geometric-grid-patterns.html)*
- *One-inch paper squares out of colored copy paper (use paper cutter to cut one-inch strips, then again to cut each strip into multiple one-inch squares)*
- *You also need triangles— make those by cutting the one-inch squares on the diagonal*
- *Glue sticks*

ACTIVITY DESCRIPTION

Each four-grid square is a potential pattern unit made from squares and/or triangles. Using only two colors, experiment and create some four-square designs—don't glue it down until you're happy with what you've made. Choose one four-square pattern to repeat four times on the larger sixteen-square grid and watch as a new, larger pattern emerges.

49

ADAPTATIONS BY AGES

Babies
Using unit blocks you can make a simple four-block grid in front of your baby. Narrate to your baby about what you are doing: "Here's the red block, and next to it let's put the yellow block..." Baby gets to handle blocks too!

Toddlers
If you're playing blocks with your toddler, why not start making your own block grid? When you've got one made, say, "Hey look what I made!" and invite your toddler to reproduce your design or make his own. If you want to make your four-square pattern with paper, invite your toddler to choose where to put each color. When that four-square pattern is completed, finish the sixteen-square together if your child is still interested. All the while talk about what you are doing and how exciting it is to see the design expand!

Older Kids
This can be a challenging activity for kids. The idea of a grid pattern unit and the process of visualizing it within the bigger grid system is often challenging enough. However, if you need even more complexity, there are ways to play with the design you choose to repeat. You can keep the shapes the same but invert the colors (for example, the color of the triangles becomes the color of the squares). You can also change the original four-square unit design so it is reflected over either the x- or y-axes, or rotated around the center point of the sixteen-square grid.

Note to the curious:
One object, such as a sixteen-square design unit, can be composed of smaller, simpler units. Simple patterns can be combined in a way that can lead to surprisingly complex designs.

Keywords:categorical variable, numerical variable, composition of units, multiples, equivalence class, pattern, transformation, symmetry, chunking, transformation, grids

CREATE YOUR OWN MATCHING GAME

Matching games are a fantastic way to develop skills in discerning similarities, sameness, and differences, even when the game is not explicitly mathematical in nature. Why not make your own mathematical matching game? There is so much noticing and conversation to be had while you trace shapes and color in your design, all mathematical in itself... and then you get to play it!

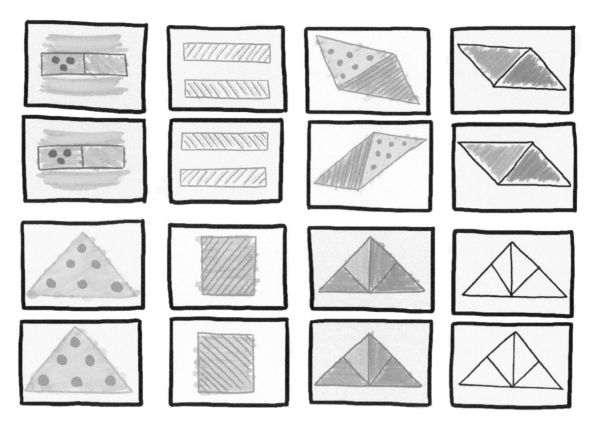

MATERIALS

- *Index cards cut in half*
- *A selection of geometric forms to trace or templates*
- *Colored pencils, crayons or markers*

ACTIVITY DESCRIPTION

Choose a shape for your first set of cards. Trace the shape onto both cards. Fill the first shape in with a color or a pattern. Create a duplicate on the second card. In subsequent designs, focus on how you are going to make each pair of cards exactly the same while also making it different in some way from every other pair. Encourage close similarities between sets to up the challenge of making the correct matches when you used the cards to play your game.

ADAPTATIONS BY AGES

Babies

Babies love peekaboo! Make four sets of cards that all use the same shape (e.g., square, triangle, circle, hexagon, etc.) but make each set a different color. Place cards face down in front of baby, turn over one card and then try to find its match. Talk it through, for example: "Is this the red circle? No! It's the blue one!" then turn the blue one face down again. When you find the color match you can say, "Peekaboo! We found the other red circle!"

Toddlers

Similar to the baby play, but this time your toddler can help you find the match. A variation would be to customize a set of cards around the one thing your child loves the most (cars, trucks, butterflies, cats, etc.) and play the game face up. For example, when Malke's daughter was two years old, Malke color copied images of different kinds of butterflies. One of each kind of butterfly was taped to a low table and the matching ones taped to popsicle sticks; her daughter loved making them "fly home" to their mommies.

Older Kids

Kids and parents can make cards together. The first game should focus on how the design or coloring of each pair of cards is different from the other pairs. You can make your matching game over the course of a week but make sure to play it when you're done! If you want to make a different matching game, go for the sneaky version! First make sure you are satisfied that your child understands the idea of difference (for example, one set is red stripes, the next is blue stripes). Then challenge your child to find a way to make two pairs of cards different from each other in very small, sneaky ways! For example, two sets of cards could use a triangle shape but include a striped design with different numbers of stripes. Take turns and see if the other can figure out the tiny differences.

Note to the curious:
Matching games are a fantastic way to develop skills in discerning similarities, sameness, and differences. If the designs are not exactly the same, what properties do they have in common? How are they different? This game also leads to discussions of congruence and equivalence.

Keywords: attributes, congruence, equivalence, properties, geometric forms, same, similar, and different

NAME CONNECTION

This is a tongue-in-cheek way to figure who should be a friend and who should be an enemy. It's on the same level of wonkiness as astrology. When they hear how they're going to decide who's a friend or not, children love to object and say that this is not the way that they make friends! How do we introduce the word "attribute" while keeping the activity light? Casually mention that attributes of names (what letters they have) are usually different from attributes of people (whether they are friends). This game makes fun of using wrong variables to sort into the friend and enemy categories.

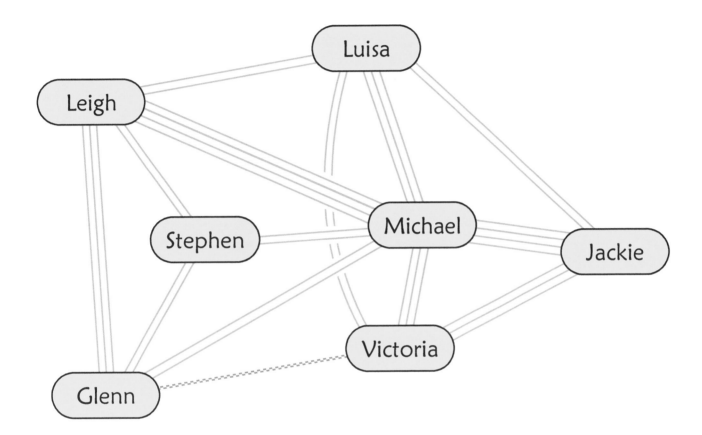

Choose a group of people you know. Write down the people's names. A group of six is good to start. If two people share at least two letters in their name connect them with green lines—one line for each letter in common. These two are friends. If two people share no letters in common—connect them with a jagged red line. These two are enemies. You succeed if connections are non-overlapping. Here are a couple of examples. The example on page 55 fails, but the example on page 56 works.

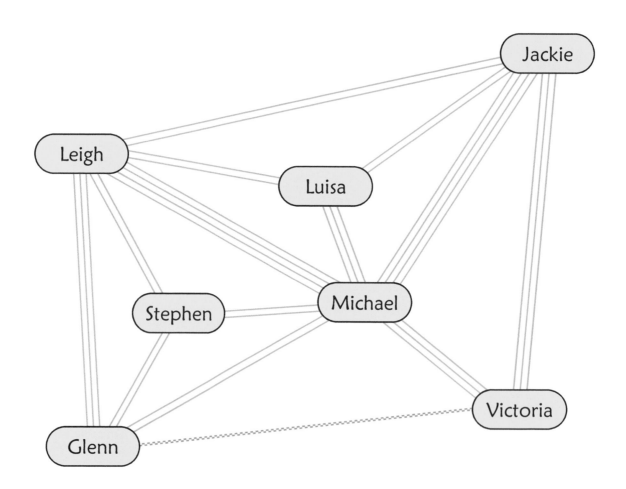

Your children will have to remember that capital and small letters are still the same letter. This is great non-mathematical practice submerged right in the middle of a math activity.

Five-year-olds and beyond usually understand the ridiculousness and humor of deciding about friendships in this way. Others may need either to use this mathematical exercise to segue into a discussion on friendships, or to sidestep the topic altogether by recasting the puzzle. Two ideas:

- What foods go well together? Chocolate and broccoli... right?
- What animals should live together in the zoo? Sharks and sparrows... right?

AROUND A STAR (SOLUTIONS)

Selfish

Fish

Caring

Bass

Sing

Octopus

Squid

October

Liquid

Fall

Spill

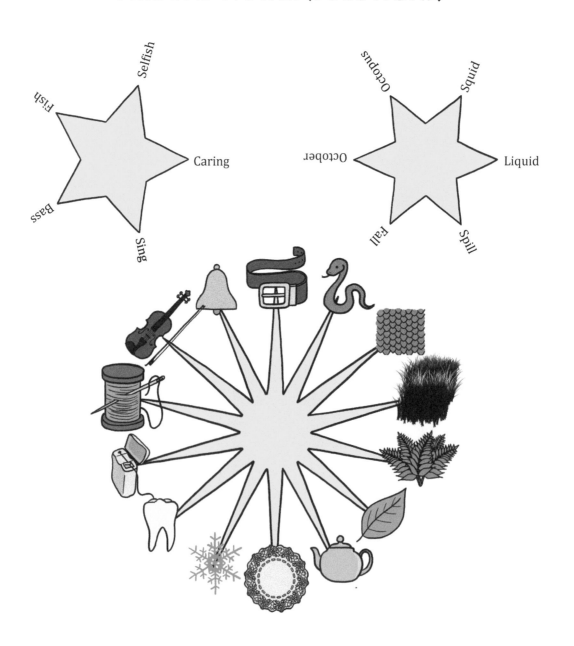

BEETLE SORT! (CONTINUED)

Collect beetles from the Curculionidae family.

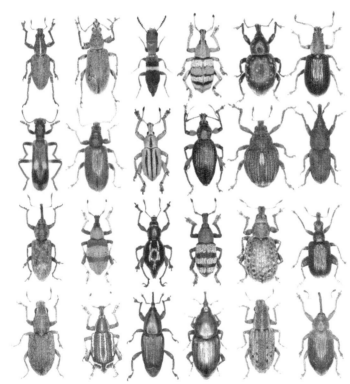

There are 2 impostors from the Cleridae family.

Collect 4 beetles from the
Leiodidae family.

Avoid those from the
Lucanidae family.

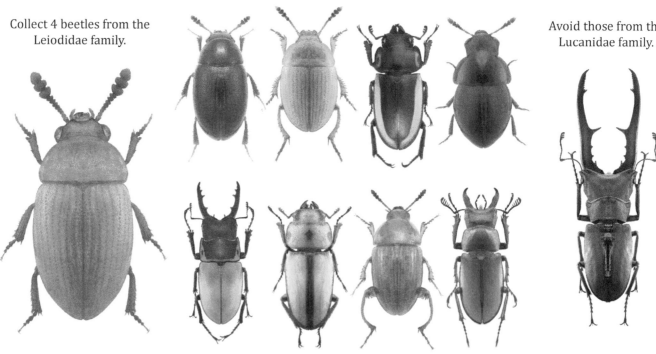

Collect 4 beetles from the
Meloidae family.

Avoid those from the
Tenebrionidae family.

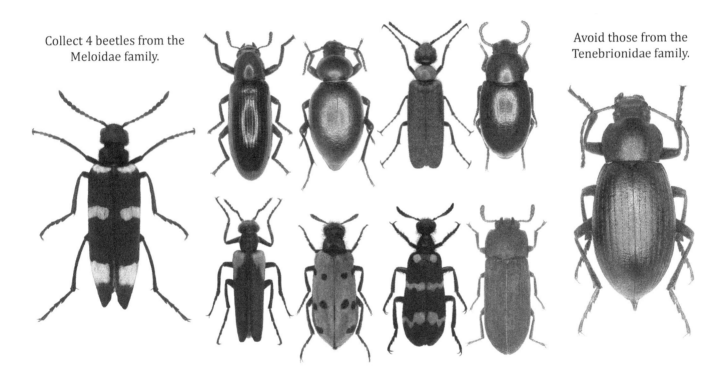

Collect beetles from the Scarabaeidae family.

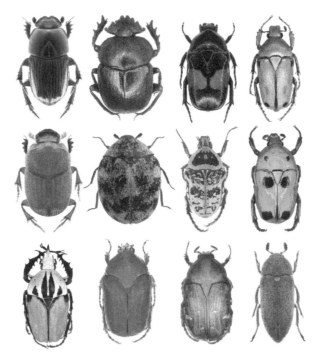

There are 2 impostors from the Dermestidae family.

Collect 3 beetles from the
Coccinellidae family.

Avoid those from the
Hydrophilidae family.

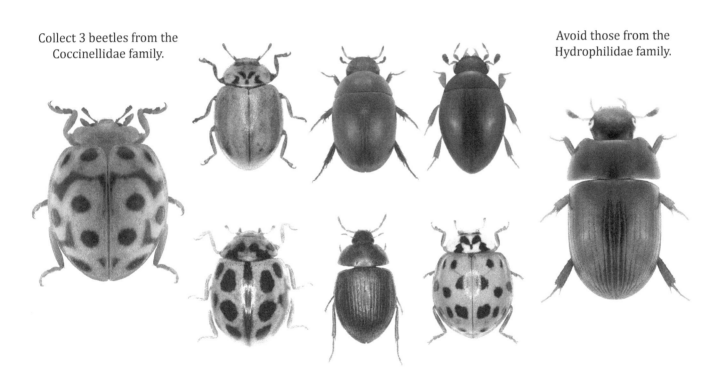

Collect beetles from the
Chrysomelidae family.

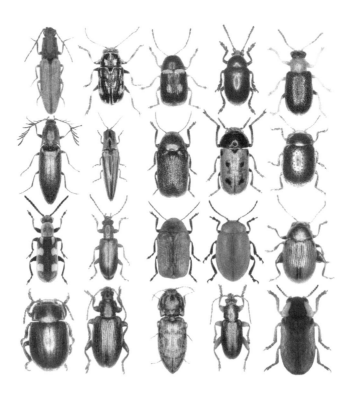

Avoid 4 impostors from
the Elateridae family.

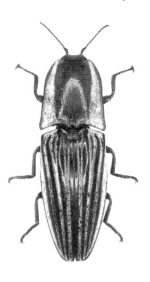

Collect beetles from the
Ptiliidae family.

There are 2 impostors from
the Trogidae family.

Collect beetles from the
Buprestidae family.

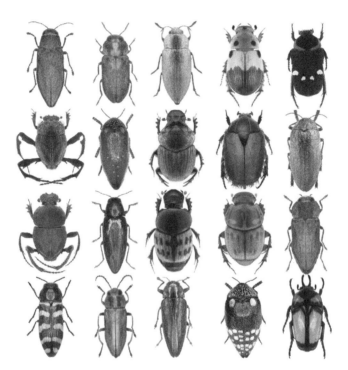

Avoid beetles of the
Scarabaeidae family.

Collect the single beetle from
the Gyrinidae family.

Avoid those from the
Dytiscidae family.

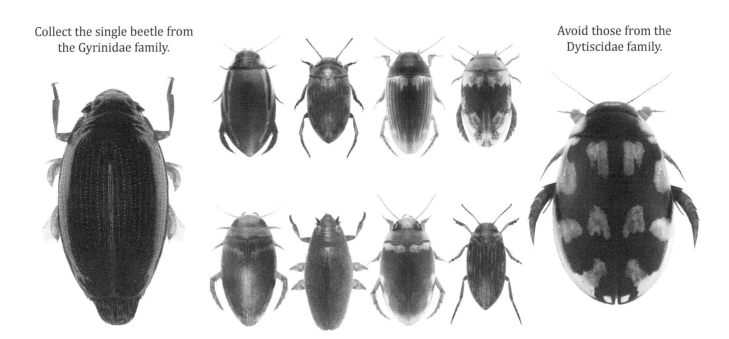

Collect beetles from the
Lycidae family.

Avoid those from the
Meloidae family.

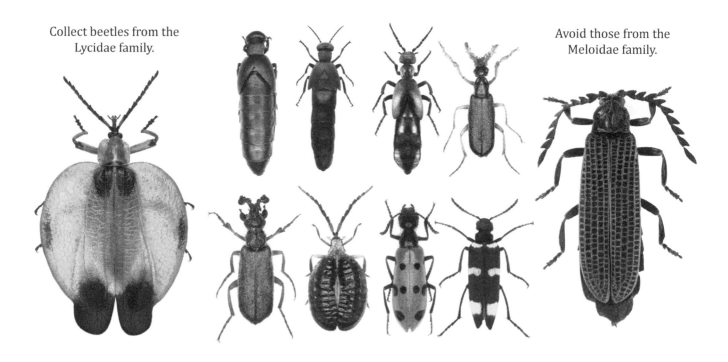

Collect 6 beetles from the
Scydmaenidae family.

Avoid beetles from the
Hydraenidae family.

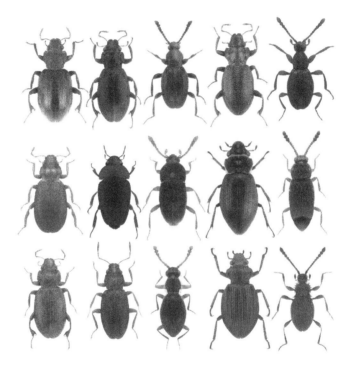

Collect beetles from the
Cerambycidae family.

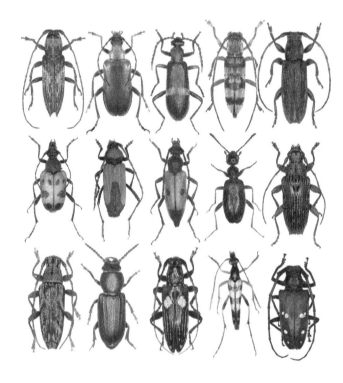

There are 2 impostors from
the Anthicidae family.

Collect beetles from the
Cerambycidae family.

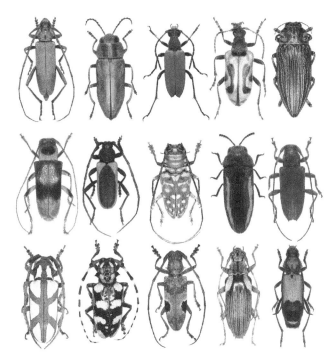

Avoid those from the
Buprestidae family.

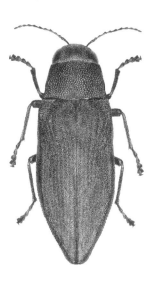

Collect beetles from the
Cerambycidae family.

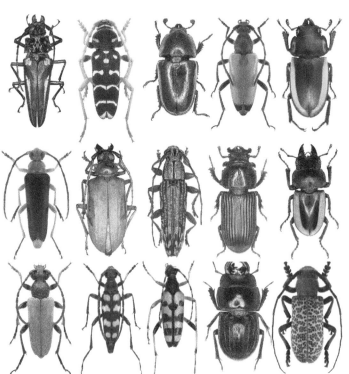

Avoid those from the
Lucanidae family.

Collect beetles from the
Dytiscidae family.

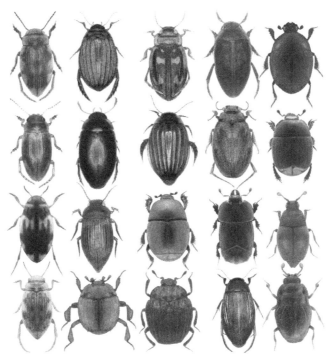

Avoid 7 impostors from the
Lucanidae family.

BEETLE SORT! SOLUTIONS

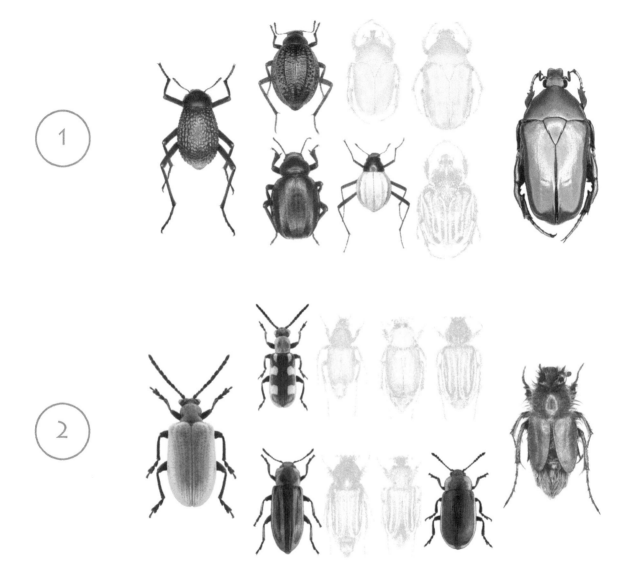

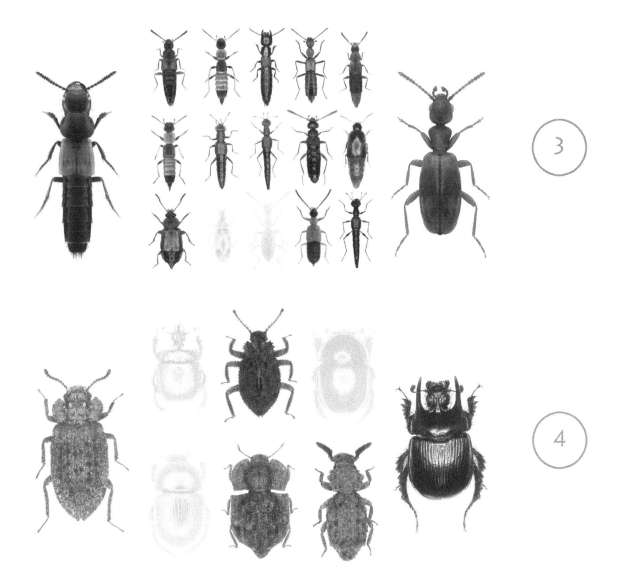

3

4

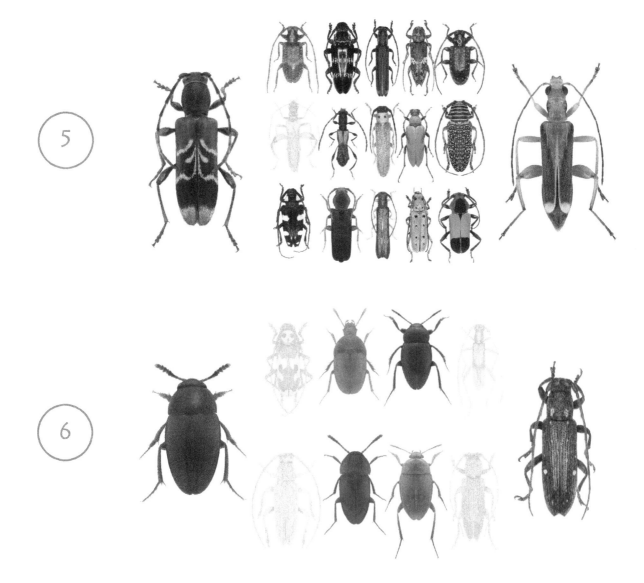

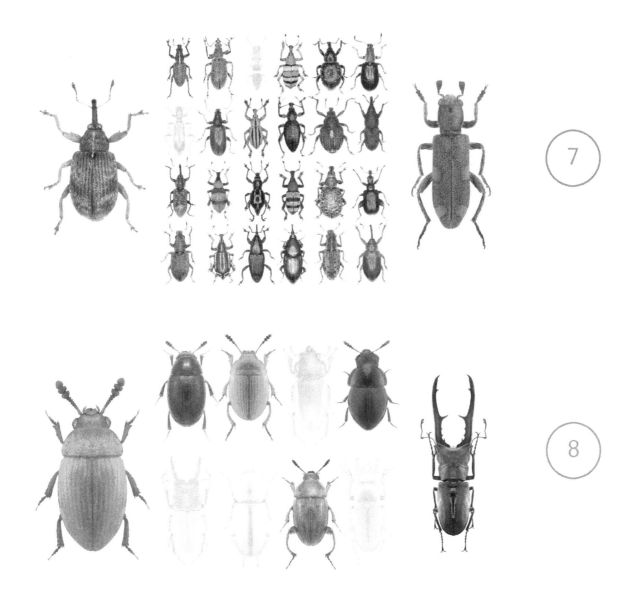

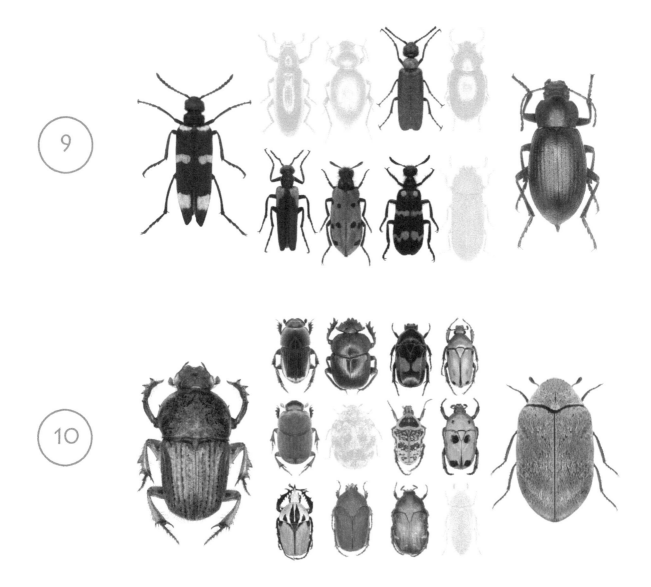

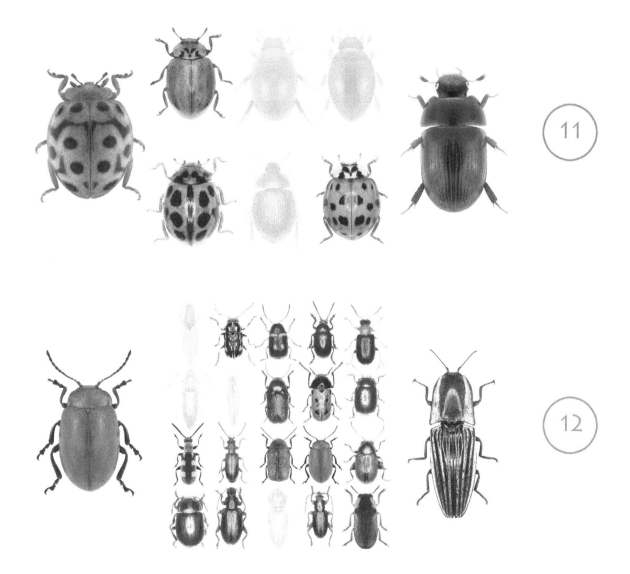

11

12

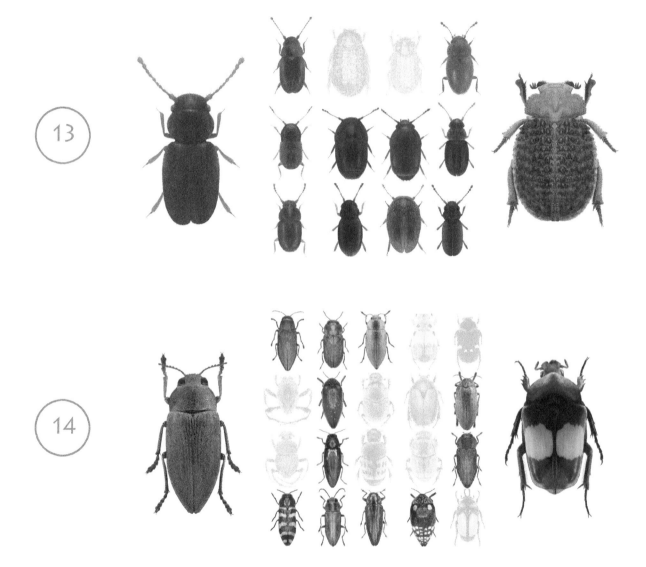

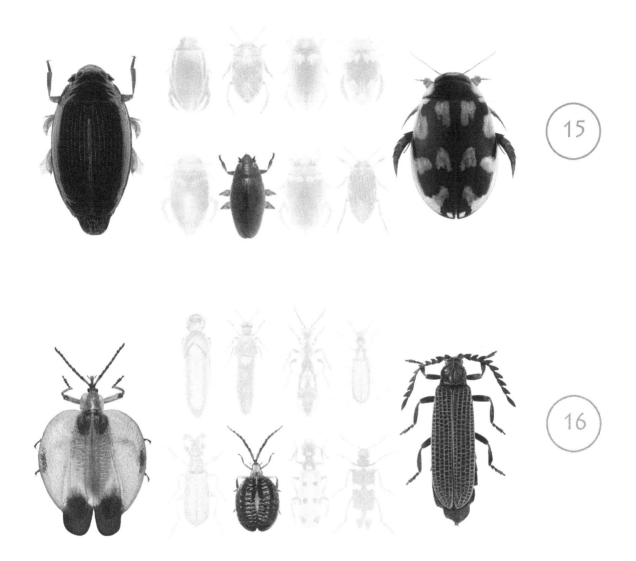

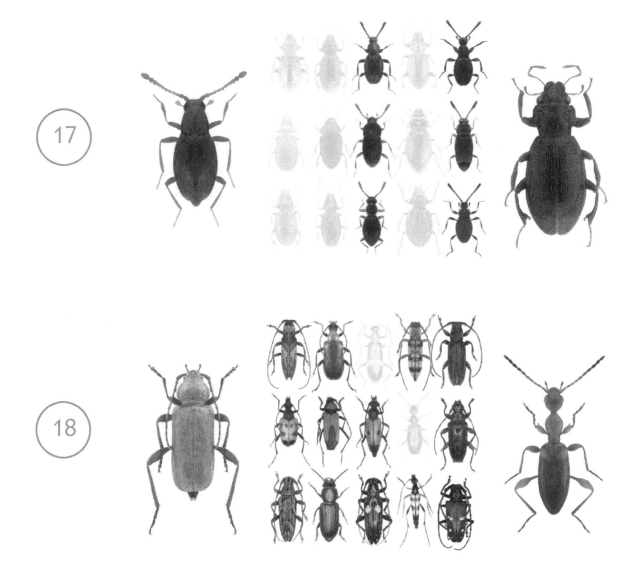

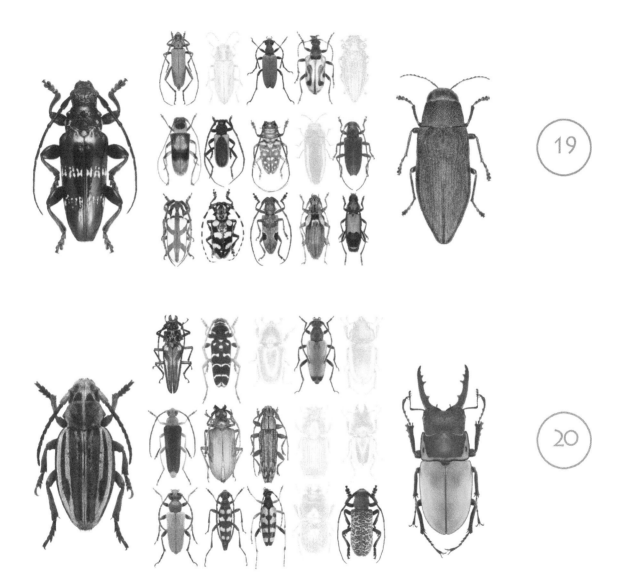

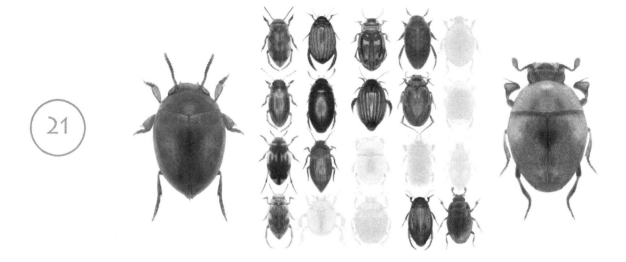

ALSO AVAILABLE

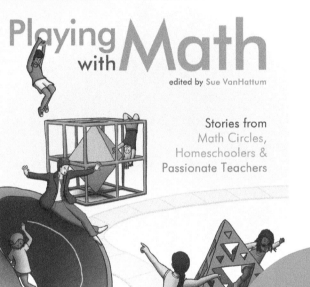

Playing with Math

edited by Sue VanHattum

Stories from
Math Circles,
Homeschoolers &
Passionate Teachers

You and your children can play with mathematics! Learn how with more than thirty authors who share their math enthusiasm with their communities, families, and students. A different puzzle, game, or activity follows each chapter to help you get started.

Available from
NaturalMath.com
and online book stores

Published by Delta Stream Media,
an imprint of Natural Math

Make math your own,
to make your own math!

A Week of Logic Games and Activities for Young People

CAMP LOGIC

By Mark Saul Ph. D. and Sian Zelbo J.D.

$A \Rightarrow B \Rightarrow C$

Available from
NaturalMath.com
and online book stores

Published by Delta Stream Media,
an imprint of Natural Math

Make math your own,
to make your own math!

This is a book for teachers, parents, math circle leaders, and anyone who nurtures the intellectual development of children. You don't need any mathematical background at all to use these activities— all you need is a willingness to dig in and work toward solving problems, even when no obvious path to a solution presents itself. The games and activities in this book give students an informal, playful introduction to the very nature of mathematics and its underlying structure.

Julia Brodsky

Bright, Brave, Open Minds:
Engaging Young Children
in Math Inquiry

Natural Math

Teach problem-solving and spark curiosity! Explore with your own children or students as you drop your own predictions and allow the children to use their tastes and ideas as a rudder. This book introduces the beginning skills of problem solving to both children and the adults who teach them.

Available from NaturalMath.com and online book stores

Published by Delta Stream Media, an imprint of Natural Math

Make math your own, to make your own math!

Avoid Hard Work!

...And Other Encouraging Mathematical Problem-Solving Tips
for the Young, the Very Young, and the Young at Heart

...UJKOVA, JAMES TANTON, AND YELENA McMANAMAN

Learn ten useful, entertaining
problem-solving techniques for
all ages. Experience problems
as explorations and inquiries
as you and your children--
even toddlers--investigate the
natural world all around you!

How do you want your child to feel about math? Relaxed, curious, eager, adventurous, and deeply connected? Then Moebius Noodles is for you. It offers advanced math activities to fit your child's personality, interests, and needs. Imagine your baby immersed in mathematics as a mother tongue spoken at home. Imagine your toddler exploring the rich world around us while absorbing the mathematics embedded in every experience. Imagine your child developing a happy familiarity with mathematics. This book helps you make these dreams come true. Can you follow your child on this journey toward a wondrous math future, without getting lost? Can you go beyond your own math limits and anxieties to guide and inspire your child? Yes! And you will not be alone. This book invites you to join a community that will answer your questions, give you ideas, and support you along the way. A five-year-old once asked us, "Who makes math?" and jumped for joy at the answer, "You!" Moebius Noodles helps you take small, immediate steps toward this sense of mathematical power. You and your child can make math your own. Together, you can make your own math.

MOEBIUS NOODLES

Adventurous math for the playground c

Available from
NaturalMath.com
and online book stores

Published by Delta Stream Media,
an imprint of Natural Math

Make math your own,
to make your own math!

ABOUT THE AUTHORS

Gordon Hamilton (Masters of Mathematics, PhD in Mathematical Biology) is a board game and puzzle designer. He founded MathPickle.com in 2010 to inject new ideas into the classroom. There is nothing he enjoys more than stumping students and having them stump him. MathPickle's primary objective is to get thirteen curricular unsolved problems into classrooms worldwide—one for each grade K-12. A conference in November 2013 established the thirteen unsolved problems. Gordon is a single father living with his two huggable children in Calgary, Alberta, Canada.

Malke Rosenfeld is a percussive dance teaching artist, math explorer, curriculum designer, editor, and writer. As creator of the Math in Your Feet program, her interdisciplinary inquiry focuses on the intersections between percussive dance and mathematics and how to best illustrate these connections for learners. She has also been highly inspired and delighted by the mathematical development of her daughter whose early years were filled with mathematical explorations, making, and play. You can read more about Malke's work at MalkeRosenfeld.com.

CPSIA information can be obtained
at www.ICGtesting.com
Printed in the USA
LVHW071015130219
607112LV00043B/706/P